The Electrochemical Society

Next Generation Photovoltaics and Photoelectrochemistry

at the 212th ECS Meeting

ECS Transactions Volume 11 No.09

October 7 – 12, 2007
Washington, DC, USA

Printed from e-media with permission by:

Curran Associates, Inc.
57 Morehouse Lane
Red Hook, NY 12571
www.proceedings.com

ISBN: 978-1-60560-182-3

Some format issues inherent in the e-media version may also appear in this print version.

Next Generation Photovoltaics and Photoelectrochemistry

ECS Transactions Volume 11 Number 9

Washington, DC., USA
7-12 October 2007

Next Generation Photovoltaics and Photoelectrochemistry

Editors:

M. Tao
University of Texas
Arlington, Texas, USA

B. Marsan
University of Quebec
Montreal, Quebec, Canada

K. Rajeshwar
University of Texas
Arlington, Texas, USA

Sponsoring Divisions:

 Energy Technology

 Physical and Analytical Electrochemistry

Published by
The Electrochemical Society
65 South Main Street, Building D
Pennington, NJ 08534-2839, USA
tel   609 737 1902
fax 609 737 2743
www.electrochem.org

ecstransactions ™

Vol. 11 No. 9

Copyright 2008 by The Electrochemical Society, Inc.
All rights reserved.

This book has been registered with Copyright Clearance Center, Inc.
For further information, please contact the Copyright Clearance Center,
Salem, Massachusetts.

Published by:

The Electrochemical Society, Inc.
65 South Main Street
Pennington, New Jersey 08534-2839, USA

Telephone 609.737.1902
Fax 609.737.2743
e-mail: ecs@electrochem.org
Web: www.electrochem.org

ISSN 1938-6737 (online)
ISSN 1938-5862 (print)

Printed in the United States of America.

Preface

The papers included in this issue of *ECS Transactions* were originally presented in the symposium "Next Generation Photovoltaics", held during the 212th meeting of The Electrochemical Society, in Washington, DC, from October 7 to 12, 2007.

***ECS Transactions*, Volume 11, Issue 9**
Next Generation Photovoltaics and Photoelectrochemistry

Table of Contents

Preface

Electrodeposition of n-type Cuprous Oxide Thin Films   1
   W. P. Siripala

Author Index   11

Facts about ECS

The Electrochemical Society (ECS) is an international, nonprofit, scientific, educational organization founded for the advancement of the theory and practice of electrochemistry, electrothermics, electronics, and allied subjects. The Society was founded in Philadelphia in 1902 and incorporated in 1930. There are currently over 7,000 scientists and engineers from more than 70 countries who hold individual membership; the Society is also supported by more than 100 corporations through Corporate Memberships.

The technical activities of the Society are carried on by Divisions. Sections of the Society have been organized in a number of cities and regions. Major international meetings of the Society are held in the spring and fall of each year. At these meetings, the Divisions and Groups hold general sessions and sponsor symposia on specialized subjects.

The Society has an active publications program that includes the following.

Journal of The Electrochemical Society — JES is the peer-reviewed leader in the field of electrochemical and solid-state science and technology. Articles are posted online as soon as they become available for publication. This archival journal is also available in a paper edition, published monthly following electronic publication.

Electrochemical and Solid-State Letters — ESL is the first and only rapid-publication electronic journal covering the same technical areas as JES. Articles are posted online as soon as they become available for publication. This peer-reviewed, archival journal is also available in a paper edition, published monthly following electronic publication. It is a joint publication of ECS and the IEEE Electron Devices Society.

Interface — *Interface* is ECS's quarterly news magazine. It provides a forum for the lively exchange of ideas and news among members of ECS and the international scientific community at large. Published online (with free access to all) and in paper, issues highlight special features on the state of electrochemical and solid-state science and technology. The paper edition is automatically sent to all ECS members.

Meeting Abstracts (formerly Extended Abstracts) — Abstracts of the technical papers presented at the spring and fall meetings of the Society are published on CD-ROM.

ECS Transactions — This online database provides access to full-text articles presented at ECS and ECS-sponsored meetings. Content is available through individual articles, or as collections of articles representing entire symposia.

Monograph Volumes — The Society sponsors the publication of hardbound monograph volumes, which provide authoritative accounts of specific topics in electrochemistry, solid-state science, and related disciplines.

For more information on these and other Society activities, visit the ECS website:

www.electrochem.org

Electrodeposition of n-type Cuprous Oxide Thin Films

W. Siripala

Department of Physics, University of Kelaniya, Kelaniya, Sri Lanka.

Cuprous oxide (Cu_2O) is generally a p-type semiconductor material due to the Cu ion vacancies exist in the crystal lattice. The technique of electrodeposition provides the possibility of depositing Cu_2O thin films resulting n-type photoresponses in solar cell devices. By maintaining pH values of the depositing aqueous acetate bath in an appropriate range, Cu_2O thin films can be potentiostatically electrodeposited. Single phase Cu_2O polycrystalline films can be obtained using this technique. The n-type behavior and the quality of the Cu_2O films produced by this technique are demonstrated by the photocurrent, capacitance-voltage, SEM and XRD measurements. Existence of oxygen vacancies in the n-Cu_2O films is evident from the photoluminescence measurements. Possibility of using these films in low cost solar cell devices is demonstrated after establishing the n-Cu_2O/p-Cu_xS hetero junction by sulphiding Cu_2O thin films.

Introduction

Cuprous oxide is an attractive semiconductor material for low cost solar energy applications (1,2). It is a non toxic, low cost and direct band gap (2 eV) material. Cu_2O is generally a p-type material due to a stoichiometry defect of copper ion vacancies (3,4). As a semiconductor, this material has been studied since early fifties, however, efficiencies of the resulted solar cells were limited to about 2% (5). It has been identified that one of the possibility to overcome this problem is to produce homojunction of Cu_2O (1,5). For this it is important to develop n-type Cu_2O material. Previous studies have shown that n-type Cu_2O films can be obtained by dipping copper electrodes in cupric ions containing aqueous electrolytes (6-8).The conditions for growing the n-Cu_2O films were the contact of Cu electrodes with weak acidic aqueous solutions. However, the resulted n-type films have very limited applications because they are limited only onto the Cu substrates. For broader applications of the n-Cu_2O films they must be able to deposit on other substrates.

We have investigated the possibility of depositing thin n-Cu_2O films on conducting substrates using the method of electrodeposition (8-11). The first step during the electrodeposition process of Cu_2O films on the electrode will be the formation of Cu+ ions at the surface due to the cathodic reaction [1] given below (12,13).

$$Cu^{++} + e^- \rightarrow Cu^+ \qquad\qquad [1]$$

The second step will be the formation of Cu_2O by reacting Cu+ with OH⁻, as given by the reaction [2] below (12,13).

$$2Cu^+ + OH^- \rightarrow Cu_2O + H^+ \qquad\qquad [2]$$

It is clear that the pH of the bath determines the formation of Cu_2O. On the other hand, it is important if there is a possibility of controlling the defect concentration of the Cu_2O film by the pH of the depositing bath and the Cu^+ concentration. This is important because Cu^+ vacancies produce p-type conductivity while oxygen vacancies could produce n-type conductivity in Cu_2O films.

In this paper it summaries the electrodeposition of cuprous oxide thin films in a weakly acidic acetate bath and the optical, electrical, morphological, structural and point defect studies of those films. Possibility of using these films in low cost solar cell devices is demonstrated after establishing the n-Cu_2O/p-Cu_xS hetero junction by sulphiding Cu_2O thin films.

Experimental

Electrodeposition of Cu_2O thin films on Ti, ITO, Pt substrates was conducted in a three-electrode electrochemical cell containing aqueous solutions of 0.1M sodium acetate and 0.01 M cupric acetate. Prior to the film deposition substrates were cleaned with detergent, diluted HNO_3 and finally with distilled water. Film thickness about 1μm can be deposited in this concentration for 45 minutes deposition duration (10). The temperature of the electrolyte was maintained at 60 ^{0}C and the electrolyte was continuously stirred using a magnetic stirrer. The counter electrode was a platinum plate and a saturated calomel electrode (SCE) was used as the reference electrode. Electrodeposition was carried out under a potentiostatic condition of -200 mV vs. SCE for 45 minutes. The pH of the electrolyte was adjusted (from 5.4 to 7.9) by adding a dilute sodium hydroxide solution or acetic acid to the bath. All the Cu_2O films were investigated in a photoelectrochemical cell containing 0.1M sodium acetate solution with a platinum counter electrode. The spectral response of the electrode was measured using a phase sensitive detection method to monitor the photocurrent signal produced by a chopped monochromatic light beam. The chopping frequency was 63 Hz. A monochromator (Sciencetech - 9010), a potentiostat (Hukoto Donko HAB-151), a lock-in amplifier (Stanford Research- SR 830 DSP), and a chopper (Stanford-SR 540) were used with a pc for the spectral response measurements. Capacitance–voltage measurements of the thin film electrodes were obtained in a three electrode electrochemical cell containing 0.1 M sodium acetate as the electrolyte and the contact area of the film with the electrolyte was 4 mm^2. The counter electrode was a platinum plate and the reference electrode was a SCE. Experimental set up consisted of a potentiostat (Hukoto Donko HAB-151), a lock-in amplifier (Stanford Research- SR 830 DSP) and a signal generator . A constant a.c. current of 1 μA at 1 kHz was passed through the working electrode and the counter electrode while keeping the desired potential of the working electrode with respect to the reference electrode. Variation of the imaginary part of the a.c. signal was measured and the capacitance (C) values were obtained at various applied potentials to get the Mott-Schottky ($1/C^2$ vs potential) plots. Scanning electron micrographs were obtained using a Philips XL 40 electron microscope and X-ray diffraction studies were performed using a Shimadzu XD-D1 X-ray diffractometer.

Results and Discussion

Electrochemical process of electrodeposition of Cu_2O can be investigated using current-potential characteristics of an inert electrode in the depositing bath. As an example, figure1 shows typical current voltage characteristics of a platinum electrode in a three electrode electrochemical cell containing 0.1 M sodium acetate with various concentrations of cupric acetate solutions, where the bath was maintained at a temperature of 55 ^{0}C.

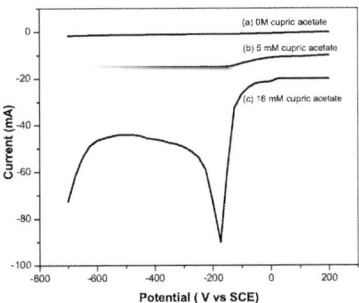

Figure 1. Current – potential characteristics of a platinum electrode in a three electrode electrochemical cell containing 0.1 M sodium acetate and cupric acetate concentrations of (a) zero (b) 0.5 mM and (c) 16 mM.

Curve (a) in figure 1 is without cupric acetate in the electrolyte. Curves (b) and (c) correspond to cupric acetate concentrations of 0.5 mM and 16 mM, respectively. In figure 1 the curves (a) and (b) have been given a zero off-set for clarity. The cathodic scans shown in fig. 1 have been started from the rest potential of the platinum electrode. The scan in pure sodium acetate solution shows no increase in current. With the introduction of Cu $^{2+}$ ions into the electrolyte a cathodic wave begins to form at 0 V vs SCE as shown in figure1 (b) and develops into a well defined peak at higher concentrations, figure 1(c). A second cathodic current rise is also evident at higher concentrations. The potential domain of the first cathodic wave gives the possible potentials for the electrodeposition of Cu_2O films. At cathodic potentials, exceeding the potential range of 0 to -300 mV vs SCE, result in the deposition of Cu.

The potential domain suitable for the electrodeposition of Cu_2O can be further verified by the X-ray diffraction (XRD) spectra obtained for the films electrodeposited at various potentials.

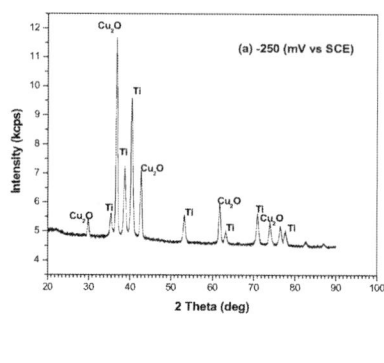

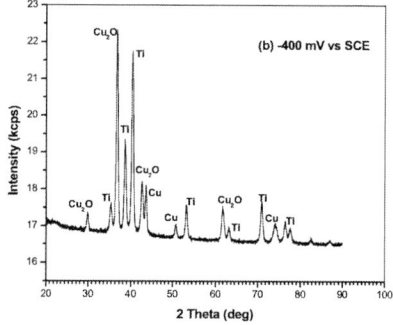

Figure 2. X-ray diffraction (XRD) spectra of films obtained by the electrodeposition at potentials of (a) -200 mV vs SCE (b) -400 mV vs SCE, on an ITO substrate. Deposition bath contained 0.1 M sodium acetate and 10 mM cupric acetate solutions at pH 6.0.

Figure 2 shows two XRD spectra of two different films deposited at (a) -200mV vs SCE and (b) -400 mV vs SCE on ITO substrates in a bath containing 0.1 M sodium acetate and 10 mM cupric acetate solution at pH 6.0. In figure 2(a) peaks corresponding to Cu_2O and ITO substrate are present. However, if the deposition potential is more cathodic Cu is co-deposited as shown in figure 2 (b). In the acetate bath Cu_2O can be electrodeposited potentiostatically on conducting substrates by selecting the potential within the potential range of 0 V to -300 mV vs SCE. Co-deposition of Cu and Cu_2O occurs if the depositing potential exceeds -300 mV vs SCE.

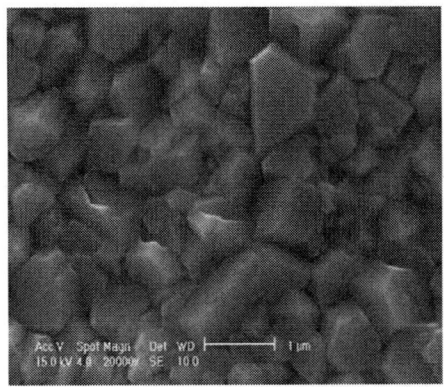

(a) Cu_2O film deposited at 55 °C

(b) Cu_2O film deposited at 20 °C

Figure 3. Scanning electron micrographs of the films obtained by the electrodeposition at potentials of -200 mV vs SCE at (a) 55°C and (b) 20°C. Deposition bath contained 0.1 M sodium acetate and 10 mM cupric acetate solutions at pH 6.0.

The scanning electron micrograph (SEM) pictures shown in figure 3 reveal the morphology of the electrodeposited films. Figure 3 (a) and (b) are the SEM pictures obtained for the films deposited at -200 mV vs SCE at (a) 55 °C and (b) 20 °C respectively. It is evident that polycrystalline films can be obtained with the electrodeposition and crystal size depends on the temperature of the bath.

The current voltage characteristics of the Cu_2O film electrode under a chopped light illumination in a PEC containing an aqueous 0.1M sodium acetate solution is shown in figure 4. The electrode was prepared by the electrodeposition in a bath at pH 6.0 containing 10 mM cupric acetate and 0.1 M sodium acetate solution. The photoresponse shows the typical behavior of an n-type film in contact with an electrolyte.

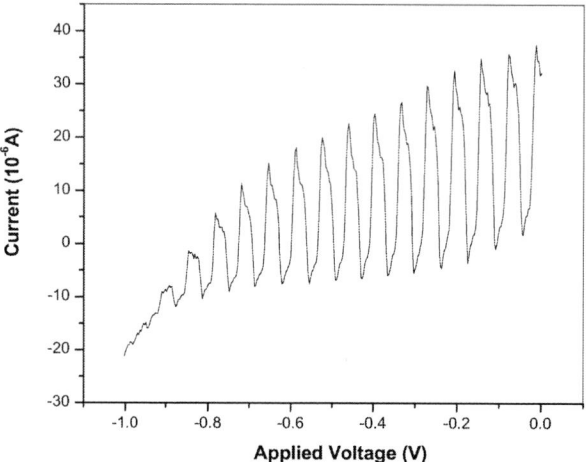

Figure 4 The current voltage characteristics of a Cu_2O film electrode under a chopped light illumination in a PEC containing an aqueous 0.1M sodium acetate solution.

The n-type behavior of the Cu_2O films is further verified by the result shown in figure 5. It shows the Mott-Schottky plot of a Cu_2O film electrode in a PEC containing a 0.1 M sodium acetate solution.

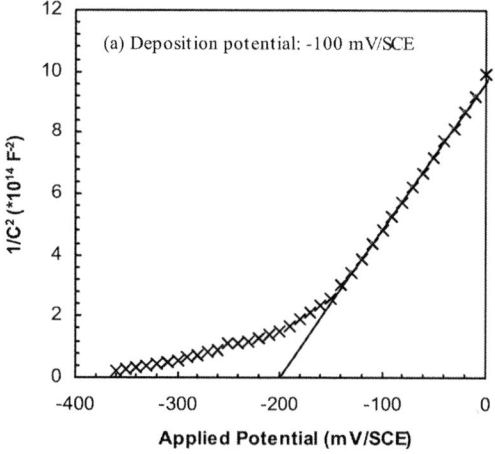

Figure 5. Mott-Schottky plot of a Cu_2O film electrode in a PEC containing a 0.1 M sodium acetate solution.

In the photoluminescence spectrum of an n- Cu_2O film measured at 80K, as shown in figure 6, additional peaks below the band edge can be observed. This observation leads to the possibility of the existence of a donor level in electrodeposited Cu_2O. This level might be originated due to the oxygen vacancies producing the n-type conductivity.

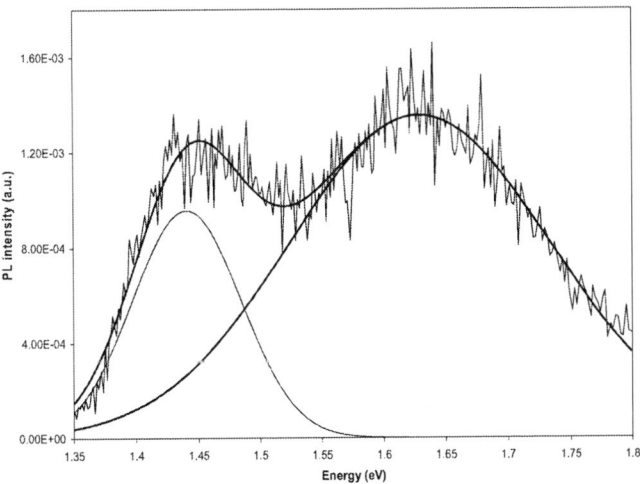

Figure 6. Photoluminescence spectrum of an n- Cu_2O film measured at 80K

It is well known that alkaline aqueous baths produce p-type Cu_2O films in the electrodeposition. We also observed that p-type films can be electrodeposited in alkaline acetate baths (14). It has already shown that homojunction Cu_2O films can be made by using the electrodeposition technique (15). Development of homojunction using the acetate bath for the electrodeposition of Cu_2O films is in progress.

We have investigated the possibility of developing a low cost solar cell with n- Cu_2O films with another low cost p-type material. For this, we have partially sulphided an n-Cu_2O film by spraying a 0.01M sodium sulfide solution over the Cu_2O film. An n-type film deposited on an ITO plate was used. Figure 7 shows the diode characteristics of the ITO/n- Cu_2O/p-CuxS/Al solar cell structure under dark and illuminated conditions. It is evident that photovoltaic characteristics are demonstrated by the n- Cu_2O /p-CuxS heterojunction.

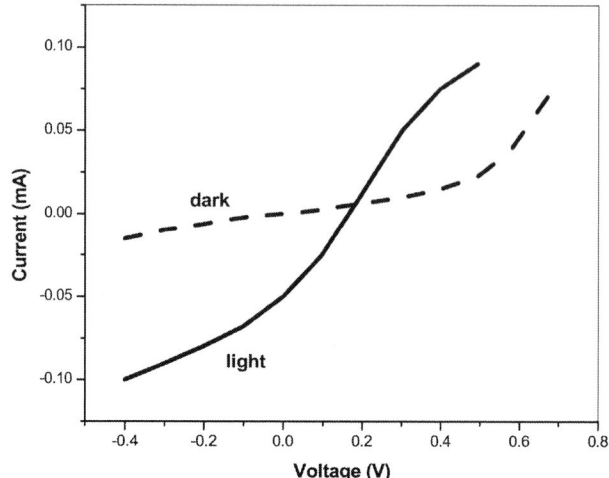

Figure 7. Dark and light current –voltage characteristics of ITO/n-Cu2O/p-CuxS/Al solar cell structure.

The photovoltaic characteristics of the ITO/n- Cu_2O /p-CuxS/Al solar cell structure were I_{SC}= 1.6 mA/cm^2 and V_{OC} = 260 mV. Comparison of the spectral response of a poor sample with a better sample is shown in figure 8. It is evident that a negative spectral response is observed in the shorter wavelengths of the ITO/n- Cu_2O /p-CuxS/Al solar cell for the structure given in (b) and, in general, for the ITO/n- Cu_2O /p-CuxS/Al solar cell structure the short wavelength response is poor. This is due to the existence of photoactive ITO/Cu_2O back contact.

.

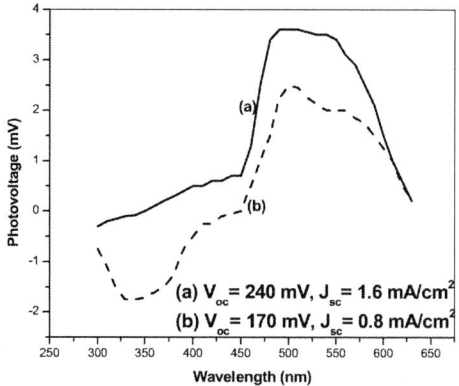

Figure 8. Spectral response of the ITO/n- Cu_2O /p-CuxS/Al solar cell structure. (a) V_{oc} = 240 mV, J_{sc} = 1.6 mA/cm^2 (b) V_{oc}= 170 mV, J_{sc}= 0.8 mA/cm^2

Obviously the above solar cell structure produces very low photovoltaic characteristics. Major problem of the structure is the existence of non ohmic ITO/ Cu_2O contact which produces an opposite photovoltage at shorter wavelengths. This opposing junction has to be eliminated. At the same time the resistivity of the films are high and therefore the photocurrents are small. In addition, there is no contribution to the photocurrent from the light absorbed in the Cu_2S layer as evident from the spectral response. If these problems can be overcome it is expected that a reasonable solar cell characteristics can be obtained with this very cheap solar cell structure.

Conclusions

It can be concluded that n-type Cu_2O thin films can be electrodeposited on conducting substrates using the method of electrodeposition. An aqueous acetate bath of weak acidic condition produces n-type films. A narrow potential domain of 300 mV is available for the electrodeposition and exceeding this potential domain result in the co-deposition of Cu. As deposited films are polycrystalline and are of good quality. The solar cell structure $ITO/n-Cu_2O/p-CuxS/Al$ obtained by sulphiding n- Cu_2O films demonstrates the possibility of developing a low cost solar cell with n- Cu_2O with other cheap p-type materials. The low cost solar cell material of Cu_2O may be able to develop to produce a reasonable efficient solar cell using the electrodeposition technique.

Acknowledgments

Prof. Rohana Garuthara, Dr.L.D.R.D. Perera , Mr. L.B.D.R. P.Wijesundara and Mr. C. Jayathillke are acknowledged for their valuable contributions. University of Kelaniya, Sri Lanka and National Science Foundation (NSF) of Sri Lanka are acknowledged for financial assistance.

References

1. B.P.Rai , *Solar Cells,* **25,** 265 (1988).
2. A.E.Rakshani, *Solid Stat. Eelctron.* **29,** 7 (1987).
3. M.Risrov, GJ Sinadinovski, M.Mitreski , *Thin Solid Films,* **167,** 309 (1988).
4. J.Bloem, *Phillips Res. Reports,* **13,** 167 (1958).
5. L.C. Olsen, F.W. Addis, W. Miller, *Solar Cells,* **7,**247 (1982).
6. S.M. Wilhem, Y. Tanizawa , L.C Yi, N. Hackerman, *Corrosion. Science,* **4,** 791 (1982).
7. F.D Quarto, S Piazza, *Electrochemica Acta,* **30(5),** 315 (1985).
8. W.Siripala , J.R.P. Jayakody , *Sol. Energy Mater.* **14,** 23(1986).
9. L.D.R.D Perera, W.Siripala , K.T.L.Silva , *J. Nati. Sci. Council Sri Lanka* **24(1** ,299 (1996).
10. W.Siripala, L.D.R.D. Perea, K.T.L. De Silva, J.K.D.S. Jayanetti, I.M.Dharmadasa, *Solar Energy Mat. and Solar Cells* **44,** 251 (1996).
11. R.P. Wijesundara , M. Hidaka, K. Koga, M.Sakai, W. Siripala, *Thin Solid Films ,***500,** 241 (2006).
12. B.Millet, C.Fiad, C.Hinnen, E.M.M. Sutter, *Corrosion Science ,***37,**1903 (1995).
13. T.Mahalingam, J.S.P. Chitra, S. Rajendran, P.J.Sebastian, *Semicon. Sci. Technol.* **17,**565 (2002).

14. KMDC Jayathileke,W Siripala and JKDS Jayanetti , *Solar Energy Mat. and Solar Cells* - submitted
15. L.Wang, M.Tao, *Electrochem. Solid State Lett* - in press

Author Index

Siripala, W. P.   1

The Electrochemical Society, Inc.
65 South Main Street
Pennington, New Jersey 08534-2839, USA

ISBN 978-1-60560-182-3